Environmental Problems in the Bakossi Landscape
A Handbook for Environmental Educators

Ekpe Inyang

Langaa Research & Publishing CIG
Mankon, Bamenda

Publisher:
Langaa *RPCIG*
Langaa Research & Publishing Common Initiative Group
P.O. Box 902 Mankon
Bamenda
North West Region
Cameroon
Langaagrp@gmail.com
www.langaa-rpcig.net

Distributed in and outside N. America by African Books Collective
orders@africanbookscollective.com
www.africanbookcollective.com

ISBN:978-9-9567172-93

Contents

"There can be no peace without equitable development; and there can be no development without sustainable management of the environment in a democratic and peaceful space."

—*Wangari Maathai,*
Nobel Peace Laureate, 2004

"Safeguarding the environment is…an essential component of poverty eradication and one of the foundations of peace and security."

—*Kofi Annan,*
United Nations Secretary-General

"The language you use with villagers, the way you relate with these people, and your general conduct within village communities are factors that determine your acceptance or un-acceptance by the communities."

—*Hans Ebako*, September 2009*
Village Councillor

*Hans Ebako hails from Mbang village in the Kupe Muanenguba Division of the South West Region of Cameroon.

Preface

The WWF Coastal Forests Programme is committed to involving Environmental Clubs of twenty target post primary institutions in the promotion of sustainable management of the biodiversity-rich resources of the Bakossi Landscape in the South West Region of Cameroon. However, such an involvement would be fruitless without equipping these clubs with the necessary tools that would enable them to face the huge challenge with confidence.

Written in simple and straightforward language, *Environmental Problems in the Bakossi Landscape* is a practical handbook aimed primarily at equipping members of the twenty Environmental Clubs with the knowledge and skills they need to be more efficient in their task of sensitising their parents and other community members and taking practical steps to address the environmental problems identified in the area.

The graphs presented in this handbook are by no means the products of a rigorous study but rather based on the data outputs of two training-of-trainers workshops organised for the target Environmental Clubs in March 2009, and can therefore be taken to present only a fair or, maybe in some instances, an unfair picture of the situation. Although the handbook focuses on the seven environmental problems of the Bakossi Landscape, its depth and breadth of analysis and the deliberate attempt at not making it too site-specific in the discussion of the problems makes it equally useful for Environmental Clubs of schools in other areas, as well as other community educators interested in environmental issues or environmental education (EE).

The handbook is structured into nine short chapters. Chapter 1 introduces the Bakossi landscape, highlighting the conservation significance of its ecological, socio-economic, and cultural landscapes. Chapters 2 to 8 are each dedicated to the analysis of one of the environmental problems, while Chapter 9 provides a brief summary and conclusion that serve to guide the educators on the appropriate areas to emphasise in their sensitisation efforts.

It is my fervent hope that every user of this handbook would be inspired to become more committed to efforts aimed at addressing the environmental problems not only of the areas where they presently live but also of any areas they find themselves, in order to make the world a better place.

Ekpe Inyang
Buea, May 2009

1

The Bakossi Landscape

1.1 Geographical and ecological description

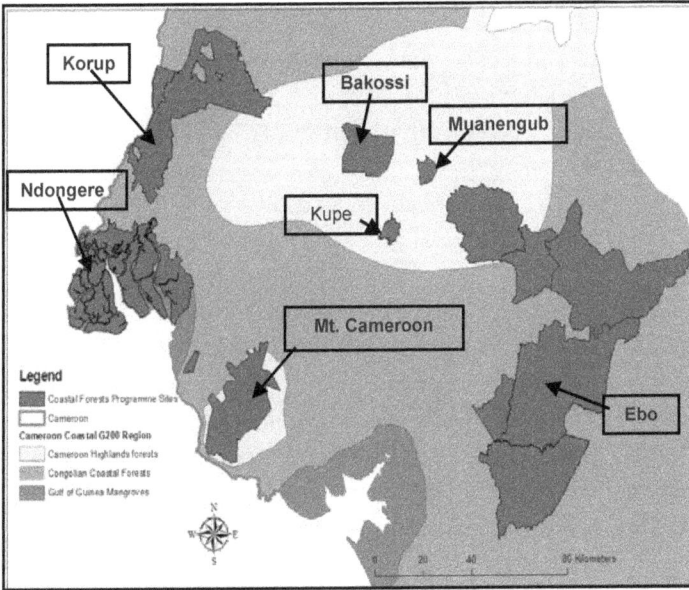

The Bakossi Landscape is located in the South West and Littoral Regions of Cameroon, precisely between latitudes 04°38'20" – 05°10'55" N and longitudes 09°22'30" – 09°58'16" E. It has a total surface area of 189,900 hectares and includes three key conservation sites: the Bakossi National Park and the proposed Mounts Kupe and Muanenguba Integral Ecological Reserves.

Washed by several streams, rivers (some of which contain waterfalls) and lakes, the landscape is characterised by a combination of highlands and lowlands that produce remarkably beautiful scenery. This combination also gives rise to a wide range of habitat and vegetation types that support different types of plants and animals. This explains why the Bakossi landscape is generally described as an area of high biodiversity and endemism, meaning that it contains different types of species, a good number of which are found no where else, such as the Mount Kupe Bush Shrike.

1.2 The inhabitants and their economic activities

The Bakossi Landscape represents a mix of indigenous and immigrant inhabitants. The indigenous inhabitants include the Bakossi, Mbo, Manehas, Bakem, and Baneka, with the Bakossi being the most highly populated and most dominant. The immigrant populations are principally the Bamiliki people who are dotted all over the Bakossi Landscape and the Bororos who are more or less restricted to the Muanenguba Mountain.

The principal economic activities are farming, hunting, petty trading, small-scale logging and livestock farming, particularly cattle rearing. The first four are mainly carried out by the indigenous inhabitants and the Bamilikis, while the last is the exclusive activity of the Bororo people.

1.3 Conservation as part of the Bakossi Landscape culture

The Bakossi Landscape represents an assemblage of indigenous communities in which conservation is traditionally at the centre of their collective culture. This can be seen from the existence of sacred forests and sacred pools, areas reserved for traditional sacrifices and other rites but where resource-depleting activities, such as farming, hunting, and fishing, are prohibited.

Another interesting area of conservation significance, when considering the culture of the people of this landscape, is the classification of certain species as taboo animals. This classification means that either female or male individuals who are indigenous to particular parts of the Bakossi Landscape are forbidden to eat some of these species, a restriction which considerably reduces pressure on them. Akin to this is the belief in totems—trees or animals believed to have spiritual relationships with certain human beings and whose death could also mean the death of these human representations. This belief often deters some hunters from hunting animals that are recognised as the most common totems, such as the Chimpanzee.

Most of Bakossi Landscape is forest

Water falls are many

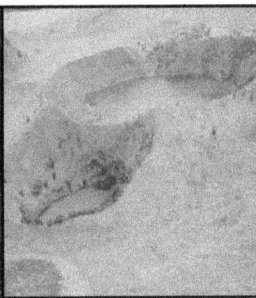

The Twin Muaneuguba Lakes

Grassland makes up quite a part of the Bakossi Landscape

Several communities have sacred forests like this one

The endemic Mount Kupe Bush Shrike

The Bakossi Landscape is wealthy, wealthy, wealthy, really wealthy! And beautiful!

The pictures here cannot give you the full picture—you need to visit the area and take a good walk around. But before then, start with these pictures. Take a look at the forest. Take a look at the grassland. Take a look at these lakes and waterfall, and this endemic bird, the Mount Kupe Bush Shrike. Take a look at the Sacred Forest with a clearing for yearly gathering for traditional stock-taking and rituals.

From your observation of these pictures, what conclusion can you draw? Are these resources not worth preserving?

3

2

Deforestation

2.1 Definition and scale

It is estimated that only 6% of the earth's surface is covered by rainforests which lie mostly within the tropics and around the equator. In spite of this small size, these forests harbour over 50% of all living species on earth (a real call for preserving them!) due particularly to their stratification into different layers, diversity of habitats, and abundance and a wide variety of food. Similarly, about two thirds of the Bakossi Landscape is covered by rainforests—lowland and montane forest types. This explains why the area is so rich in biodiversity, which translates into the availability of so many different kinds of forest products that are useful to the Bakossi Landscape people in particular and mankind in general.

The rainforests are also important in other ways. They protect the soil from direct heat of the sun and from the impact of rain drops. The roots of the trees hold the soil together, prevent rapid flow of water, and thus reduce erosion. The forests also protect water catchments, thus making the water in streams, rivers, and lakes available all year round. They are often described as "the lungs of the earth" because they take up carbon dioxide and send out oxygen, thus facilitating the cycling of these gases. They sequestrate or store excess carbon dioxide in tree trunks and, therefore, control the amount in the atmosphere that would have otherwise increased the greenhouse effect that contributes to global warming. The forests also help in maintaining the water cycle, and their transpiration function helps to regulate local climates by the cooling effect of the moisture in the surrounding air.

Today the forests of the Bakossi Landscape, like those of most areas around the globe, can no longer perform their functions to the maximum, due to a serious event known as deforestation. Deforestation can be defined simply as the cutting down of the trees that make up the forests at rates that far exceed those of replanting and natural regeneration. When this occurs it signals, to a large extent, the disappearance of forests or their replacement with other vegetation types. This is the most severe problem, as we shall see later, and one that is common to all the forested areas of the Bakossi Landscape, particularly the Kupe and Bakossi sites.

Scale of Deforestation in the Bakossi Landscape

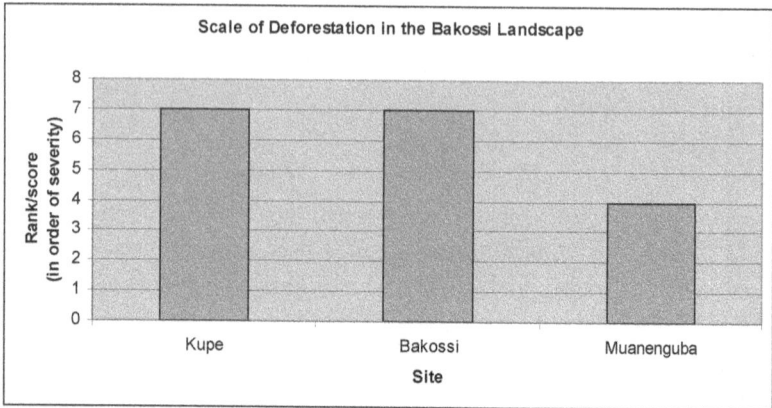

2.2 Causes and underlying issues
The causes of deforestation in the Bakossi Landscape, as in many parts of Africa and elsewhere, are shifting cultivation, logging for timber, felling for fuelwood, growth and expansion of human settlements, and establishment of pasture lands.

Shifting cultivation, characterised by what is popularly known as slash-and-burn agriculture, is the main cause of deforestation. It is a traditional system of farming in which the farmer cuts down and burns new tracts of forest every year to establish new farms. Often all the trees in a cleared portion are cut down to make it easy to do the burning, to ensure the availability of fuelwood, and also to ensure that there is enough sunlight for the crops. The situation can best be described as "scrambling for land", as farmers are in a sort of competition to acquire new farmlands. This rush for land is fuelled by perceived scarcity highlighted by population growth and uneven distribution of fertile soils. Prestige seems to be another motivating factor, as each farmer wants to be recognised as hard-working by the amount of land he or she has acquired. Poverty cannot escape mention here, as a majority of the farmers strives to make reasonable incomes by establishing large cash crop farms or plantations.

Similar to the shifting agriculture, though not as widespread and intense, is clearing of forest tracts to establish pasture lands for the grazing of livestock such as cattle. The cleared areas are often burnt to stimulate the growth of grass and this converts the portions into

6

permanent grass fields. In the Bakossi Landscape, this is practised exclusively by the non-indigenous Bororo people who are now more or less permanent inhabitants of the Muanenguba site.

Logging for timber is another major cause of deforestation. This is done at two levels: small-scale, domestic tree felling using chainsaws and large-scale, commercial logging using bulldozers. The former is carried out by local and non-resident people for income and domestic consumption, often without any licence or authorisation from the government, while the latter is carried out by logging companies operating for huge financial gains, with licences from and taxes to the government. Although the local communities know that most of the tax money goes to the government, a majority of them strongly support these logging companies in order to gain access roads, as well as secure some employment for community members.

Road construction in itself is also a cause of deforestation, as many kilometres of road mean destruction of a huge number of trees and other plants. This is similar to establishments of human settlements in which large areas of forests are destroyed for building and construction, especially where such settlements are large and continue to expand as the population grows.

Ignorance, poverty and even greed are some of the underlying factors that seem to drive the activities that cause deforestation. Let us take the case of ignorance, for example. Even though people may observe that deforestation is taking place due to their farming and other operations, they do not seem to consider the phenomenon as having any future negative consequences. Instead, to them this may simply mean that the community is developing, as closeness of forests to communities is often mistakenly regarded as a sign of underdevelopment. Similarly, many people believe that it is only by cutting down most of, or all, the trees in their farms that the crops they grow would do well; they do not know that some crops actually do better under tree shades.

Even when the consequences eventually start manifesting, some people may still find it difficult to link these with deforestation. Of course, they seem to be blinded by the effects of poverty, to the extent that they tend to consider the forests as a resource that must be exploited to whatever extent to make a living. Greed is not absent as a factor in the strong desire to constantly cut down the forests, as some people often strive to acquire more than others.

2.3 Impacts and potential solutions

Deforestation is already having far-reaching impacts on some communities in all the three sites of the Bakossi Landscape. These impacts include scarcity of some forest products, extinction of some species, soil erosion, land degradation, water pollution, water scarcity, and increase in local temperature.

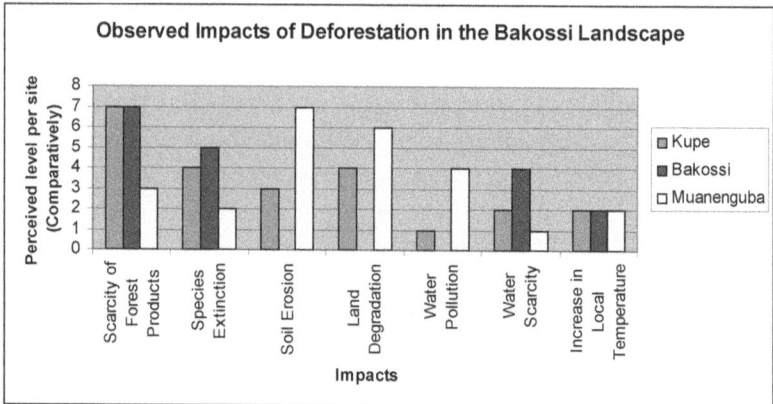

When huge tracts of forests are heavily logged for timber or expansive areas are cut down and burnt for agriculture, many forest products soon become scarce due not only to intentional or accidental damage of individual organisms but also to the destruction of key habitat areas. This situation, if not checked, could lead to the extinction of some species, especially those that are endemic or restricted to the affected areas or which cannot find appropriate habitats in nearby areas. It should be noted that deforestation often causes what is known as habitat fragmentation or split into smaller patches (as there is often encroachment into the same key habitat area at different parts) which causes the area to look like small islands. This situation often results in the separation and reduction in the sizes and structures of animal populations, to the extent that the new collections of individuals are unable to contribute to the growth of their populations. In the same vein, continued burning under shifting cultivation means that original plant species can no longer re-colonise their areas of origin and may be completely eradicated.

Deforestation also means that there is exposure of the land to direct heat of the sun and to soil erosion, which eventually results in the degradation of the land. This could be seen from the deformed nature of some parts of the landscape, as well as the compactness, hardness, and unproductiveness of soils in some areas, often manifested in the stunted and scanty nature of the vegetation that still grows on it. Water pollution occurs as erosion washes rubbish, fertilizers, and other chemicals, particularly from agricultural systems, to streams, rivers, and lakes, causing huge losses of aquatic life and high incidences of water-borne diseases (This will be discussed in more detail in Chapter 7).

Water scarcity sets in especially where deforestation occurs along streams and rivers, around lakes, and/or in water catchments. In Chapter 8 we will learn more about the role that trees perform in protecting water bodies and their catchments. Meanwhile, let us consider temperature increase as another consequence.

Local as well as global temperatures are increasing particularly because of the increasing amounts of carbon dioxide in the atmosphere. Carbon dioxide and other greenhouse gases form a thick blanket that traps and reflects the heat radiated from the sun back to the earth, thus causing the temperature to increase. We have already learnt that the trees take up excess carbon dioxide from the atmosphere and store this in their trunks. And when the huge numbers are cut down and burnt or allowed to decay, the carbon dioxide is released back into the atmosphere. As the load of carbon dioxide in the atmosphere from this and other sources becomes too heavy for the remaining trees and other plants to control, the temperature increase could become even more noticeable.

We have seen that the impacts of deforestation are huge, and these often translate into extreme poverty and misery. Should we then sit and wait for a miracle? Certainly not. How can we address the situation? The most direct way to tackle the problem is by encouraging the communities to carry out intensive reforestation; that is replanting of many trees in the deforested and degraded areas. And since ignorance is one of the issues that are seen to fuel the problem, it would be wise to start by thoroughly educating the communities on how the observed impacts are linked to their activities, and what they could do to reverse or ameliorate the situation.

Education should include aspects of land use planning and land tenure, with particular emphasis on the necessity for water catchments protection. Similarly, farmers should be advised to do selective felling when opening new farms and to gather the destroyed biomass in heaps before doing the burning, where this cannot be totally avoided, in order to reduce the total surface area affected by fire. They should also be encouraged to practice agro-forestry. Community leaders should equally be encouraged to take the necessary steps to ensure that the advice is taken seriously. There should also be discussion on the importance of increasing the use of improved biomass stoves, in order to cut down on the rate of fuelwood consumption as a factor of deforestation in the Bakossi Landscape. Presently only very few people, mostly those who sell what could be termed fast food near drinking spots and other strategic places, use at least the less efficient models of this type of stoves.

The forests are cut down so fast

Source: Google

Improved biomass stoves (like this type on the right) retain more heat, reduce the quantity of fuelwood required to cook food, and considerably reduce the amount of carbon dioxide emitted

Bush burning is a common agricultural practice in the Bakossi Landscape

Structure of a logged forest showing persistent gaps after several years

Let's plant more and more trees to address this problem

Do we need the forests? These forests are disappearing rapidly, due particularly to shifting cultivation (characterised by slash-and-burn agriculture) and logging.

Every year each farmer cuts down a new portion of forest. If each new portion, to be modest, is half a hectare, and there are fifty serious farmers in a village, how many hectares are destroyed each year by that village?

How many of such farming communities are there in the Bakossi Landscape? And at that rate of destruction, how many hectares of forest are destroyed each year in this landscape?

Logging, too, is taking a toll!

Is this a good or bad picture? If you think it is a bad picture, then we need to reduce the rate of deforestation in the Bakossi Landscape, by practising agro-forestry and employing other land management methods or techniques. For areas that are already deforested, we should consider doing intensive replanting of trees there.

3

Soil Erosion

3.1 Definition and scale

The soil can be described as a thin layer that covers the land or earth's surface. It is the main and natural store of nutrients that support plant growth, including the crops we grow for reliable food supplies. This means that both animals and we humans depend principally, though indirectly, on the soil, for without it we will not have food to eat.

Generally speaking, the soil is only a thin, outermost portion of the land. In some areas, particularly in dry or arid regions such as deserts, it is hard to say whether there is anything there that merits the name of soil at all, since what exists scarcely supports plant growth. This is due particularly to the significantly low humus content, as well as other essential elements. Even in some rainforest areas, such as we are in, the layer can be very, very thin, and is said to be fragile by virtue of its susceptibility to erosive forces and other forms of destruction.

By contrast, however, the forests of the Bakossi Landscape are blessed with fertile soils which support a wide variety of vegetation types and promote high agricultural productivity. This exception is possibly due to past volcanic activities in the area, which ejected and deposited huge masses of soil-forming and soil-enriching materials on the land.

It is regrettable, however, that the "volcanic blessing" of this landscape is gradually fading away, due to pronounced soil erosion in some areas. Soil erosion can be defined as the movement of soil mostly by water and wind, but sometimes through landslides caused by the effect of gravity. In the Bakossi Landscape, water and landslides are the main agents of soil erosion. The soil is carried from one part of the land and deposited on another part, especially in valleys, including those that serve as channels for streams and rivers. Soil erosion is said to be severe when it occurs at a rate faster than that of soil formation and rejuvenation. In the Bakossi Landscape, with the exception of the Bakossi site, the situation is apparently becoming more and more serious in the Kupe and Muanenguba sites, particularly the latter whose main vegetation is grass.

Scale of Soil Erosion in the Bakossi Landscape

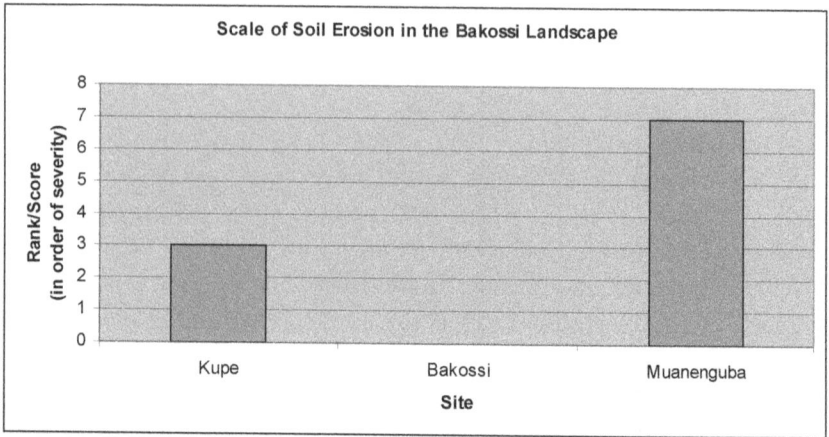

3.2 Causes and underlying issues

We have already learnt that erosion occurs where the soil is exposed to the effect of, say, running water. This often happens "silently" naturally, but the rate is more pronounced due to human activities. The main causes of erosion in the Bakossi Landscape include deforestation, overgrazing, and poor farming methods.

Deforestation and overgrazing are the common means by which the soil is exposed to erosion. In the case of deforestation, our main concern here is farming. The rate at which erosion occurs is higher when farming or overgrazing takes place on steep hill slopes. In the case of farming, poor farming methods also add to the intensity of the problem. For example, where farmers make their ridges vertical to the hill the current of the runoff is usually so strong along the ridge channels that a lot of soil on the slope is washed down into the lowland areas or valleys.

Ignorance seems to play a vital role in the promotion of soil erosion. To begin with, most farmers do not appreciate the severity of the impact of this problem on their agricultural effort. Even when there is crop failure they tend to attribute this to natural or other causes rather than to erosion. Since erosion often occurs slowly and unnoticeably, especially at first signs, they do not see how this can seriously affect their farms in terms of productivity. Also, they often do not see the link between their activities and the occurrence of erosion.

3.3 Impacts and potential solutions

The already observed impacts of soil erosion in all but the Bakossi site are land degradation, soil infertility, and water pollution.

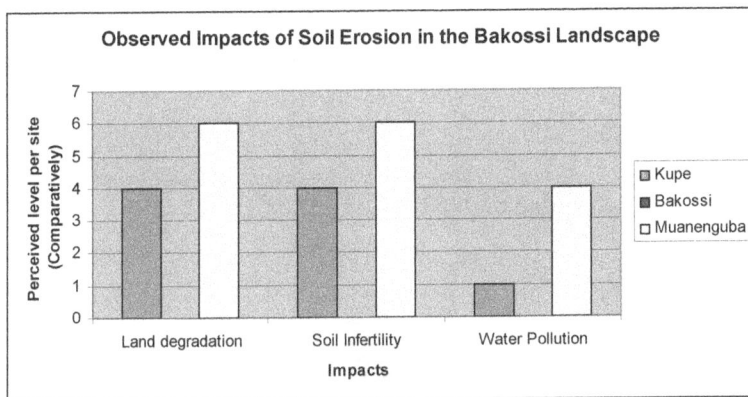

Observed Impacts of Soil Erosion in the Bakossi Landscape

We have already discussed the processes of land degradation, soil infertility, and water pollution in Chapter 2, under impacts of deforestation. We can only reiterate here that land degradation has negative repercussions particularly on the natural vegetation and scenic beauty. Soil infertility simply portends food insecurity, famine, and extreme poverty for the communities. Water pollution means increase in the incidences of water-borne diseases. The present level of the observed impacts should serve as a warning also to the Bakossi site communities and should push people to take appropriate measures to address the problem of soil erosion in the Bakossi Landscape.

The most direct way of tackling the problem is to embark on an aggressive campaign for the planting of appropriate trees or grasses in degraded and erosion-affected land areas. This is to ensure that the soil is held together by the roots of the plants which would additionally serve to reduce the erosive force of the runoff. The inclusion of leguminous tree species would additionally help to increase the fertility of the soil through the process of nitrogen fixation by special bacteria in the nodules found on the roots of these plants.

Again, it would be expedient to start by educating the communities on the causes of soil erosion and the gravity of the impacts of this problem, especially relating to agricultural productivity and health. They should equally be warned of the possibility of losing most of the essential components of their soils—the plant nutrients—to other land areas where they might never gain access to carry out their farming operations, even though this might take several years to happen.

There is already serious erosion on most of the hills in the Bakossi Landscape...

During the rainy season runoff from heavy downpours transport our important plant nutrients away to other communities, or into valleys, rivers, and lakes, as we continue to expose the soil by cutting down trees and clearing the grass. This means that we are losing the force behind our agricultural productivity.

We can check erosion by reducing our rate of cutting down the forests, replanting trees in areas where we have destroyed a lot, and avoiding farming on hill slopes. Let us not sit and watch soil erosion take away our natural agricultural input.

...Yet people continue to farm on the hills

4

Land Degradation

4.1 Definition and scale

The land can be described simply as the outer part of the earth's crust that contains the soil and all that it supports. We cannot, therefore, talk about the soil without talking about the land, because there can be no soil without land. But unlike the soil that can be moved about by the process of erosion, the land in itself is immobile. Furthermore, it contains far more than just the soil, as in some areas it includes several kinds of minerals that can be found on it or buried and hidden deep, deep in it. Land is, without doubt, an important habitat for humans and several other living species.

The land is an important asset for any human community, and the Bakossi Landscape people are no exception. It has many different uses, including building, agriculture, and recreation. It is also an important source of income. For instance, many families in the Bakossi Landscape depend on the sale of some portions of their land to raise reasonable amounts of money. In fact, land has become such an important commodity of trade, as many people now struggle to secure not only land for building but also arable land in response to perceived scarcity due particularly to observed degradation.

Land degradation can be defined specifically as the destruction of the biological potential of the land, notably its ability to support plant growth. This implies that land degradation, although a broader concept, is often linked to soil degradation or soil impoverishment. In this respect, land degradation is said to be taking place when an arable piece of land is showing signs of sterility or unproductiveness. However, land degradation does not confine itself to a reduction in the arable quality of land only; it includes a reduction in the aesthetic quality of the land as well. The Kupe and Muanenguba sites are already experiencing drastic reductions in the arable and aesthetic qualities of some areas of their land, with Muanenguba being the most affected.

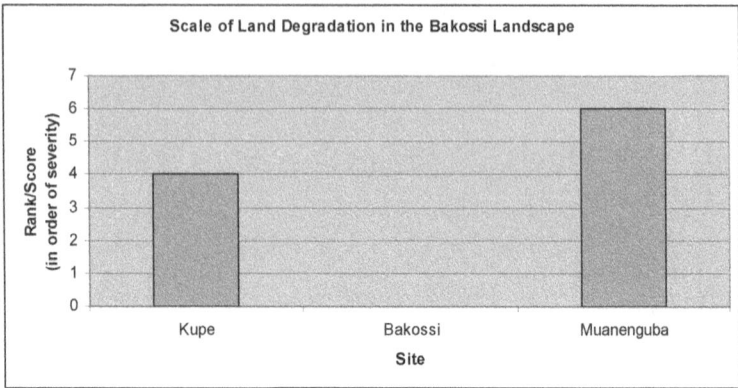

Scale of Land Degradation in the Bakossi Landscape

4.2 Causes and underlying issues

We have noted that land degradation is defined in terms not only of its arable quality or fertility but also its aesthetic quality or natural beauty. But what can affect these qualities? Deforestation, logging, erosion, pollution, overgrazing, and poor farming methods are known to be the major factors that contribute significantly to the relatively high rate of land degradation, particularly in the Kupe and Muanenguba sites.

Deforestation and logging damage the structure and natural beauty of the forest and expose the soil to agents of denudation and erosion. This also applies to overgrazing and poor farming methods, such as over-cropping and weeding without mulching, which render the soil susceptible to erosion. Erosion not only hastens the depletion of soil nutrients but also deforms the land by creating gullies, potholes, and other such ugly features.

Bush fires also expose the soil to erosion, affect its texture and directly cause some plant nutrients to escape from it in the form of gases, and for soil-forming organisms to be destroyed, due to the excessive heat of the fire. Similarly, pollution with solid wastes not only destroys the natural beauty of the land but also makes it difficult to be put it into effective use or renders it completely useless. On the other hand, chemical pollution largely affects the fertility of the soil, particularly by rendering it acidic.

Ignorance obviously plays a significantly important part in land degradation. Like erosion, land degradation may occur so gradually that people simply do not consider it to be able to have any serious negative consequences in the foreseeable future.

Of course, even when the consequences are foreseen, or are already experienced and are clearly linked, without doubt, to certain

18

current human activities, poverty seems to push people to continue doing what they have started doing, just to make a living. In fact, in poverty situations, people simply tend to damn the consequences, which may instead plunge them into more poverty.

4.3 Impacts and potential solutions

The impacts of land degradation so far observed, notably in the Kupe and Muanenguba sites, include loss of natural vegetation, destruction of nature's beauty, soil infertility, and poor harvests or low agricultural yields. Starting with loss of natural vegetation, it is regrettable that a landscape known for its unrivalled diversity of vegetation types is already showing signs of losing this natural gift in some parts. This invariably means loss of potential medicinal and other important species of the Bakossi Landscape.

Linked to the loss of natural vegetation is loss also of nature's scenic beauty that characterises the Bakossi Landscape. To some people, this aspect of natural beauty as a natural resource may appear inconsequential. But consider the huge losses if an area that is already attracting such an increasing tourist attention suddenly becomes something like a wasteland in which no one takes interest any more. It may become a subject of heated debate if some people are told also that nature's beauty has a lot to do with lifting our spirits, and even contributes to good health. Loss of soil fertility and the concomitant poor harvests are also manifesting in some communities of the affected sites. These have implications for food security and, especially, income earnings in the affected communities, as most people depend largely on agricultural production.

Education is of paramount importance in addressing the problem of land degradation. People need to be sufficiently educated and sensitised on the causes and impacts of this problem, as well as encouraged to take practical measures to address the situation. For example, since deforestation is one of the causes of land degradation a practical way of addressing the problem is to carry out tree planting campaigns, with focus particularly on the planting of agro-forestry or leguminous species. While these trees help check erosion as a cause of the problem, they also add to the natural beauty and contribute to soil fertility. Proper application of organic manure would also help to enrich the soil.

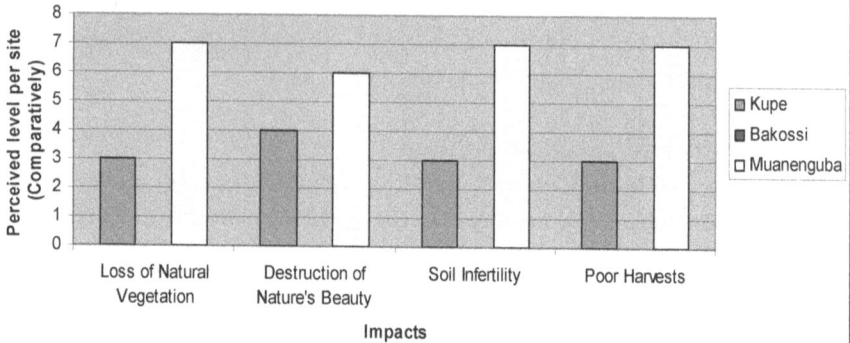

Observed Impacts of Land Degradation in the Bakossi Landscape

Y-axis: Perceived level per site (Comparatively) — scale 0 to 8
X-axis categories: Loss of Natural Vegetation, Destruction of Nature's Beauty, Soil Infertility, Poor Harvests
Legend: Kupe, Bakossi, Muanenguba
X-axis label: Impacts

The issue of over-grazing could be addressed by discouraging the free-range method, involving the leading of cattle from one open-access area to another, and encouraging the creation of individual or small group pastures in more or less confined areas. This ensures that individuals or group members are made more careful in their grazing methods and manage their pastures sustainably. For instance, this new method might make them consider the number of cows each should keep in a confined pasture, considering its size, in order to avoid a tragic situation where the animals start dying out because the available grass is not at all sufficient to meet their food requirements.

This is a frightful picture we do not want to see in the Bakossi Landscape, do we ?

Source: Cunningham, William P., Saigo, Barbara Woodworth and Cunningham, Mary Ann. 2003. *Environmental Science: A Global Concern.* New York: McGraw-Hill.

But look above at what is already happening here, especially in the Muanenguba site. If the vulnerable hills continue to be cultivated and exposed to the severe soil erosion characteristic of the site, the now distant, frightful picture on the left will become no distant possibility here.

Land degradation is a sign that there is something seriously wrong. Not only is Nature's beauty lost but it might be indication also that the soil has been seriously depleted or is rapidly being exhausted. If nothing is done about it as soon as the situation is observed, it might become very difficult, if not impossible, to bring the land back to its original, arable state, as shown in the frightful picture on page 26.

The piece of land behind these environmental club members is slowly becoming degraded

Establishing tree nurseries to support tree planting is a first step towards addressing land degradation

5

Wildlife Scarcity

5.1 Definition and scale

Wildlife refers to both the plants and animals that are found in the wild, although the term is often applied, erroneously, only to the animals. By the term wild here is meant natural areas, such as forests, grasslands, streams, rivers, lakes, and so on.

Wildlife, that is the plants and animals, perform a range of ecological functions. For example, trees in the forest not only protect the soil and water catchments but also facilitate the cycling of carbon dioxide and oxygen, sequestrate excess carbon dioxide in their trunks and provide habitat and food for the animals. The animals, on their part, are of great importance to the plants by pollinating their flowers and dispersing their seeds.

Many rural communities in Africa, as elsewhere—the Bakossi Landscape communities being no exception—depend almost entirely on wildlife for survival. For example, the different kinds of plants and animals found in the Bakossi Landscape are a source not only of food, medicine, building materials and fuelwood but also income for the communities. Some of the plants and animals are also a major source of attraction for researchers and tourists; thus their promotion of scientific research and eco-tourism which could bring in additional income for some communities.

Unfortunately, some of these natural resources are increasingly becoming scarce in all the three sites of the Bakossi Landscape. Wildlife scarcity describes a situation where some important plants or animals are in short supply, and thus can no longer satisfy the demands or interests of the local people, as well as researchers and tourists that visit the area. This, in other words, means that some of the animals or plants are hard to find because of their considerably reduced numbers. This situation seems to be more serious in the Kupe and Bakossi sites than in the Muanenguba site.

Scale of Wildlife Scarcity in the Bakossi Landscape

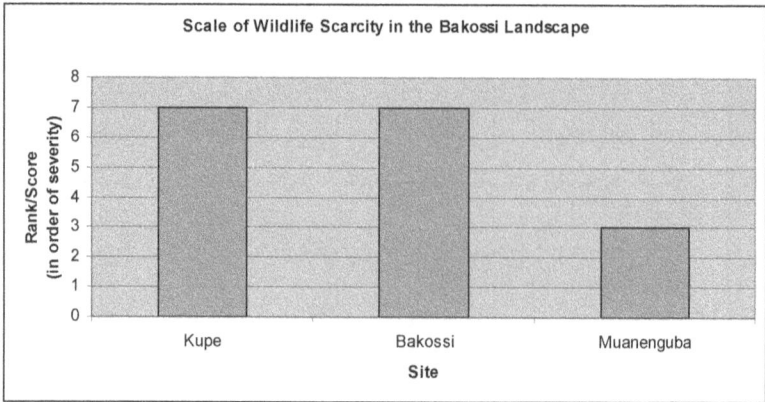

5.2 Causes and underlying issues

Habitat destruction, over-hunting, trapping, over-harvesting, and poor harvesting methods are identified as the major causes of wildlife scarcity. Habitat destruction is a silent and indirect way of destroying wildlife, as a lot is often lost without notice. This happens because, in the case of animals, these creatures are not only deprived of where to live but what to eat. Normally, the animals in a destroyed habitat area may try to escape to find territory and food reserves in similar but distant areas in the course of which they may either starve to death, get seriously injured or killed by other animals, or encounter and killed by hunters on the way.

Over-hunting is a direct way of reducing the populations of animals. It is common in some communities to find young people who have dedicated themselves to full-time hunting, spending several days in a single trip to the forest hunting animals. In some communities, foreign hunters are allowed in to swell the number, thus making the situation even worse. The foreign hunters may spend even more time—often up to a month!—in a single hunting trip than the indigenous hunters who may switch to other activities, including responding to community work or attending some important family or community meeting. You may find these people returning with huge sacs of Bushmeat everyday or after every week. This practice helps to deplete the resource very fast.

Trapping is another direct way of reducing animal populations. It is even worse than hunting, as a lot of the animals may even get rotten by the time the trapper goes round searching the often numerous traps he has set. Further to this is the fact that traps do not discriminate, and may catch animals that are too young to be fit even

24

for household consumption. Trapping may, therefore, turn out to be a destructive or wasteful activity.

Just as with over-hunting and trapping, over-harvesting and poor or non-selective harvesting of timber and non-timber forest products could eventually make these resources become scarce within the shortest time. Take, for example, the ring-barking method that some people use in harvesting Prunus. If you removed the bark all round the trunk of a particular stand, do you expect to meet it still alive when you return to the area, say, after six months? This is the situation with the harvesting of many other non-timber forest products, in which some people harvest by cutting the vines or stems right at the base, thereby reducing their chances of regenerating.

Another interesting example of poor harvesting methods is the use of small-mesh-size nets, capable of entangling from the biggest to the smallest fish, employing techniques like scooping, casting, crossing, and trawling. Often the tiny fish are left behind to rot, when badly injured or dead, as they are often too tiny to be used for food. When still alive, they are thrown back into the stream, river or lake where some may sooner or later die from injuries sustained from the entanglement in the net. The use of small-mesh-size nets, like using gamalin (see Chapter 7), is a rather wasteful method that depletes the fish resources in no time.

It is hard to say that people employ the above unsustainable methods and practices out of ignorance, but this seems to be the case. For example, do these people know that what they are doing could result in scarcity in the future—a situation which would, in turn, increase their level of poverty and suffering? Of course, poverty has been forcing them to employ the above methods and practices, just to earn a living.

Perhaps, commercialisation of the products is a more serious driving force, as high demands mean that more and more exploitation has to be carried out within the shortest time possible and using the most effective, albeit unsustainable, methods available—just to satisfy the customers! In fact, there are rich people out there who have such strong appetite for some of the products (for example, bushmeat and timber) that they are ready to pay anything to get these. Similarly, the traditional use of animal trophies, such as elephant tusks and leopard skins, as symbol of power and authority has been responsible for the increased scarcity of certain animal species in many areas.

Inadequate or lack of alternative protein and income sources, such as cane rat and snail farming that produce two of the most important delicacies, must have also contributed significantly to the observed wildlife scarcity in some parts of the Bakossi Landscape, due to increased pressure on the limited number of available sources.

5.3 Impacts and potential solutions

Increased time to find wildlife resources, and poverty are the main impacts of wildlife scarcity in the Bakossi Landscape.

These impacts are felt in all the sites, and people now spend a lot of time looking for some important wildlife resources due to considerably reduced abundances or increased distances to get to areas where some can still be found. In this situation only those who are strong and can trek long distances are able to harvest such resources. Also, a lot of energy and time that would have been invested in other productive activities are spent just to fetch very small quantities. Since people in the rural communities depend almost entirely on wildlife or forest resources to earn a living, scarcity obviously has serious economic implications, with increasing poverty becoming the order of the day.

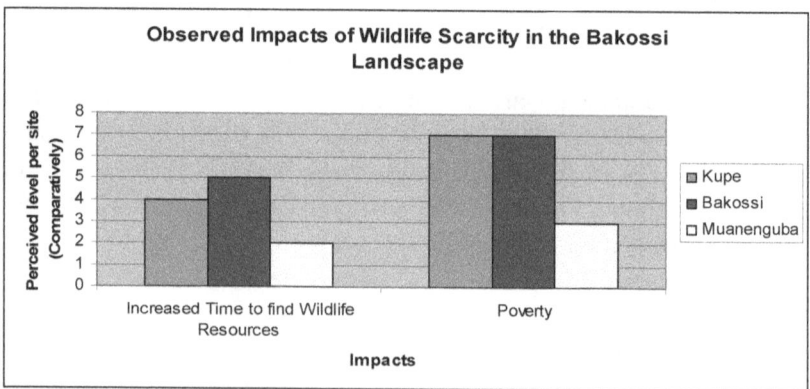

To practically address the problem of wildlife scarcity, there is need for the promotion of alternative protein and income generating activities, such as cane rat farming, snail farming, bee farming, and domestication of wild plant species, such as *eru* and Prunus. In fact, it would be wise to consider domesticating all the scarce, economically important wildlife resources of the area, except where it is forbidden by the law to do so.

Education on the causes of wildlife scarcity and the observed impacts is of paramount importance. With this, people might eventually see the need to reduce or manage the exploitation of their resources, and to embark on the development of alternative sources. Where the expected results are taking too long to arrive (notably when many people still do not desist from employing unsustainable methods or exploiting already threatened resources) before the resources are pushed to the brink of extinction, it would not be considered harsh to encourage community leaders to call in the government for rigorous law enforcement.

It is important to note that the situation (that is, wildlife scarcity) is so glaring and alarming in some communities that some community leaders and other concerned individuals are already pushing strongly for the application of this method, in order to bring it under control. A more efficient and, perhaps, milder approach would be to encourage the formation of what has come to be known as Village Forest Management Committees. Fully recognised and supported by the government, these committees operate for and on behalf of their respective communities to regulate the exploitation of wildlife resources.

Ring-barking kills the tree   **Bushmeat displayed for sale**

Eru (*Gnetum africanum*) is exported
in huge quantities almost every week

Bushmeat pepper soup

Cane rat farming can bring in huge incomes

Snail farming generates income too

Domesticate eru and make real cash

If we continue to exploit our natural resources as rapidly (and using such unsustainable methods) as we presently do they will soon become much scarcer than they are at the moment.

When our natural resources are scarce, we become poor and when they are very scarce we become very poor. This means we depend on these resources and should use them with great care.

With increases in population, we should seriously consider reducing our pressure on these resources; otherwise our children will meet nothing physical but only see these things in pictures or hear about them in stories.

We can considerably reduce pressure on the natural resources by doing eru farming, snail farming, and cane rat farming.

6

Species Extinction

6.1 Definition and scale

A species of plants or animals is a group whose individuals can interbreed and reproduce fertile offspring; that is, young individuals that are capable of reproducing also when they reach maturity. For example, the species blue duiker, known locally as *frutambo*, can only reproduce fertile, young blue duikers when individuals of the blue duiker species interbreed. In other words, it is only blue duiker individuals—males and females—that can contribute to the population growth and sustainability of the blue duiker species.

About 1.8 million species on earth (plants and animals combined) are said to have been named so far. But many continue to be discovered and classified, indicating that there are more species on earth than we know presently. This also means that most of the species may even become extinct before they are discovered, in other words, before their uses are known.

Species extinction can be defined as the disappearance of a species from the earth as a whole or from its geographical region of origin. We refer to the complete disappearance of a species from the earth as global extinction and that which is restricted to a particular geographical region as local extinction. It should, however, be noted that what we might consider as local extinction could actually be global extinction, for example, when this affects endemic species, species found only in a particular geographical region and nowhere else. This situation is due to the fact that we might not even know what species are endemic to the regions in which we live. In the Bakossi Landscape, an area known for its high degree of endemism, species extinction has been reported in all the sites, with the Bakossi site being the hardest hit, followed by Kupe. For example, a number of important animal species that once sprawled the landscape have not been seen for years, such as the leopard, giant pangolin, chimpanzee, elephant, and buffalo.

31

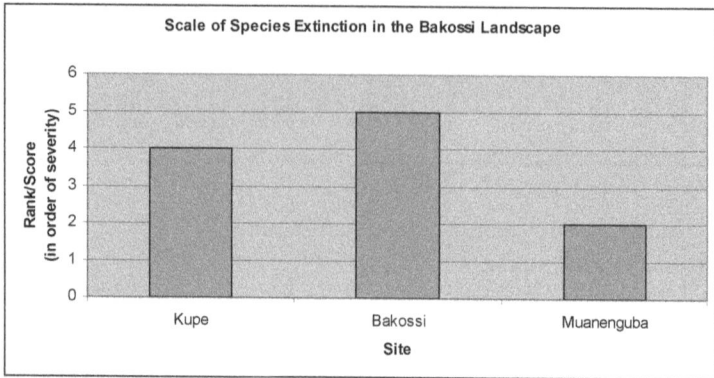

Scale of Species Extinction in the Bakossi Landscape

Rank/Score (in order of severity) — y-axis: 0 to 6
Site — x-axis: Kupe, Bakossi, Muanenguba

6.2 Causes and underlying issues

Over and above habitat destruction, over-hunting, trapping, over-harvesting, and poor harvesting methods that have already been discussed under wildlife scarcity, bush burning and pollution are also the known causes of species extinction in the Bakossi Landscape.

Once the wildlife resources are already scarce, due to habitat destruction, over-hunting, trapping, over-harvesting or poor harvesting methods, this could be an indication that they are endangered or threatened with extinction. At this point if these causes are not addressed, this unfortunate, irreversible phenomenon might occur.

Bush burning and pollution might not give any time for any warning to be received, but result in the complete extermination of a species within a very short time. This is so because the incidence of bush fire and pollution, especially at devastating and intolerable levels, is like a war situation in which species could die in huge numbers almost at once. Traditional use of trophies, as discussed in Chapter 5, could also drive species to extinction, especially with increases in the number of people who receive titles annually.

Where there is sufficient warning through observed wildlife scarcity and people still do not cease employing the above unsustainable and destructive methods, this could simply mean that they are either ignorant about the possible consequences or that they simply do not care about their future and the future of their children.

6.3 Impacts and potential solutions

The obvious impacts of species extinction are loss of biodiversity, loss of traditional knowledge, and poverty.

Biodiversity can be defined as the variety or different kinds of plants and animals available in an area. It can be described as our natural bank containing different accounts that serve our various needs. Biodiversity loss, to use the above analogy, is like the systematic closure of the accounts, which means our needs are increasingly facing the problem of not being satisfied. This invariably signals poverty and hardship, due to the loss of our sources of livelihood. Biodiversity loss also means a great loss in traditional knowledge which is an important part of the culture particularly of rural and indigenous peoples.

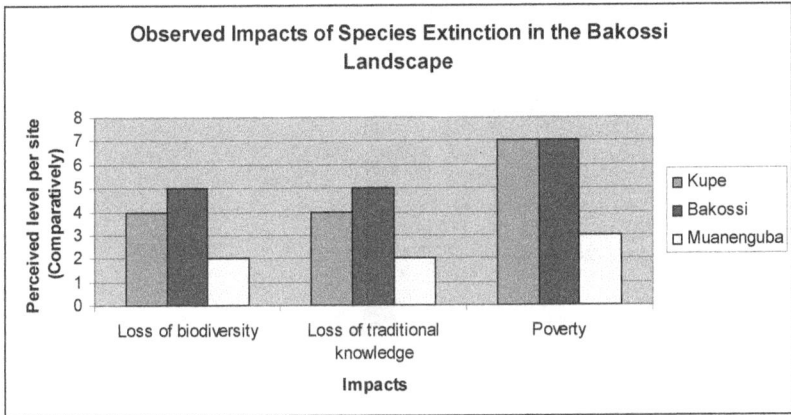

Observed Impacts of Species Extinction in the Bakossi Landscape

Addressing species extinction requires more than just educating people on the rate of occurrence and gravity of the problem; it is expedient also to employ the stricter method of law enforcement. Otherwise, many more species might become extinct before we realise what is happening.

To address species extinction due to traditional use of trophies, it might be necessary to promote the use of "fake" rather than "real" trophies, such as wood carvings of elephant tusks (as already practised in some Bakossi Landscape communities) and woven materials that resemble, say, leopard skins. This would require educating the users and at the same time encouraging those skilled in producing the above craftworks.

33

The potto is endangered

Many animals are already endangered, i.e. in danger of becoming extinct, and are totally protected by law. These include, (for mammals): elephant, leopard, chimpanzee, water chevrotain, tree hyrax, potto, and giant pangolin, (for birds): Mount Kupe Bush Shrike, brown-necked parrot, white-throated mountain babbler, and Bannerman's weaver, and (for reptiles): crocodile and goliath toad.

If there is still pressure on these animals, they will become extinct and our children and our children's children will not have the opportunity also of knowing them, except by pictures.

And do we want to leave a "picture world" or a real world for our children and our children's children?

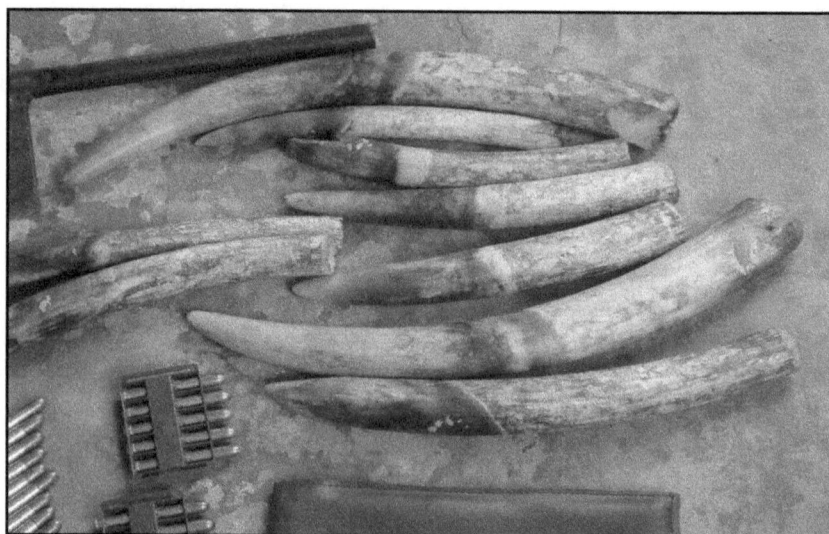

Elephants are killed mostly for their tusks

7

Water Pollution

7.1 Definition and scale

Water is an important resource that every living organism cannot do without. Scientists estimate that about 75% of our bodies are made of water. Similarly, water takes up about 75% of the earth's surface, of which 97% of all the water makes up the oceans. Two thirds of the remaining 3% is locked up in glacial ice, and subtracting underground water leaves us with only 0.008% that forms the lakes, rivers, and streams. It should be noted that some of the underground water may find its way to the surface as springs that flow into valleys to form streams and rivers or collect in some depressions to form lakes. The streams normally form tributaries of rivers which eventually empty themselves into the oceans.

There is generally the impression that water is so abundant, especially where there are big rivers, lakes, and oceans. True, there may be plenty of water, but the question we must ask is if it is available for drinking and other domestic uses. It should be noted that water from the oceans, and even some rivers, is salt water that is unsafe for drinking not only because of its characteristically high salt content but also because of its equally high level of pollution. Even some fresh water streams, rivers, and lakes may also be so seriously polluted that their water is not safe for drinking. Water is said to be polluted when it is contaminated or contains substances or micro-organisms, known as pollutants, in quantities that make it unsafe.

We learnt in Chapter 1 that the Bakossi Landscape is washed by streams, rivers, and lakes. However, some of the streams and rivers, and even lakes, are gradually becoming polluted. While water pollution seems to be serious in the Muanenguba site, the Kupe site (with the exception of the Bakossi site that does not seem to be facing the problem yet) seems to be only now experiencing some level of the problem. But pollution being a silent but serious killer, this picture presents an early warning also for the Bakossi site.

Scale of Water Pollution in the Bakossi Landscape

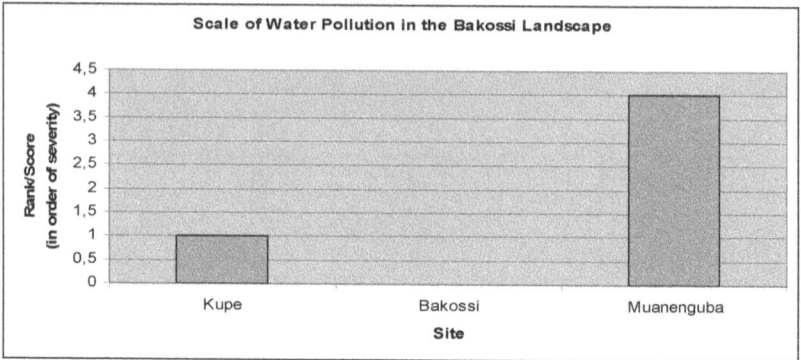

7.2 Causes and underlying issues

Poor sewage and domestic waste disposal methods, inappropriate application of fertilizers, use of poisons to kill fish, and contamination of the water catchments are some of the obvious causes of water pollution in the Bakossi Landscape.

In some communities in the Bakossi Landscape, as in some rural communities in Africa and elsewhere, people have the habit of dumping wastes and even defecating in streams and rivers, thereby contaminating or polluting the water. It should not come as a surprise that some rural homes that live near the banks of streams or rivers and own water system toilets might simply connect the drainage pipes direct into these water bodies, obviously in order to avoid the task of having to occasionally empty the sceptic tanks.

It is important to note that fertilizers, though useful for soil enrichment, is actually a pollutant when it joins the water system. Therefore, inappropriate use of fertilizers, for example, application near river courses or on erosion-prone hill slopes is a source of pollution for some rivers, streams, and lakes. Also, over-application of fertilizers, particularly in areas prone to leaching, could result in the contamination of the underground water. By the same token, when water catchments are contaminated in this or any other way, it is but obvious that the streams and rivers that take their rise there are also polluted. The use of pesticides, such as gamalin, to kill fish is a very dangerous way of polluting streams, rivers, and lakes.

Is it that some people do not know that they are contributing to water pollution by carrying out the above activities? Is it that they do not appreciate the impact of water pollution at all? Or is it simply a demonstration of irresponsibility or "I don't care" attitude on the part of some people when they do these things?

Ignorance may be playing a part in the case of wrong application of fertilizers, because many farmers may not even know that fertilizers are a potential source of water pollution. However, we cannot use ignorance as an excuse when it comes to people defecating in streams or rivers; this should simply be seen as a demonstration of irresponsibility or "I don't care" attitude by such people. What about pouring poisonous chemicals into these water bodies as a method of fishing? It is clear that some people would want to make quick money by whatever means, especially when they are desperately hit by poverty. But this could also mean that such people are simply careless or ignorant about the harm they themselves or their family members are exposed to when they use such dangerous methods.

7.3 Impacts and potential solutions

The obvious impacts of water pollution are high incidences of water-borne diseases, water scarcity, rapid destruction of aquatic resources, and species extinction. We have already noted that water pollution is due to soil erosion particularly from agricultural systems.

The rubbish or dirt washed or dumped into streams, rivers, and lakes stimulates the growth of germs that may cause serious outbreaks of water-borne diseases. This creates a terrible and frustrating situation where water is available but scarce, since it is not safe for drinking or other domestic uses.

The misapplied fertilizers that would have accompanied the rubbish into the water bodies help to enrich these. This situation, known as cultural enrichment or eutrophication, promotes algae bloom or rapid multiplication of algae. The increased algae population soon uses up the dissolved oxygen very rapidly, thereby depriving other aquatic life of this very important resource. This phenomenon is referred to as biological oxygen demand. The inevitable consequence of this is rapid decline in aquatic resources due to increased mortality or death rates.

The effects of chemical pollution could be even more serious, destroying aquatic life in large numbers and affecting people or animals that feed on the poisoned aquatic resources. Some of the chemicals are resistant and non-biodegradable, remaining in the environment for long periods of time. Eventually these recalcitrant chemicals become part of the food chain, accumulating and increasing in concentrations, as you go up the food pyramid, in the tissues of the species that make up the various trophic levels—a process known as biomagnification.

The concentrations may reach such high levels that when finally we eat, say, the big fish or other animals that live on the small, "polluted" fish, we seriously suffer the effects of the chemicals. This could manifest in serious health problems or in the sterility of some people due to the "chemical poisoning" which, in our communities, is often blamed on witchcraft. This is often the situation with other animal species, which often silently leads to their extinction.

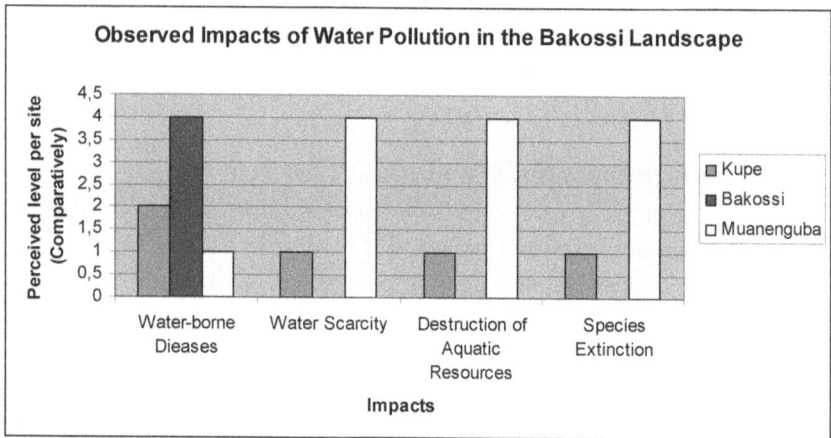

Addressing the problem of water pollution requires a lot of education and sensitisation since many people seem not to know the serious impacts linked to this problem. This can be seen from the popular saying that "an African does not die of dirt" (but certainly always dies of witchcraft!). To erase this false notion, it might be necessary to start by carrying out environmental sanitation campaigns and then encouraging the creation of Health Committees (where they do not exist) and establishment of community waste dumps in isolated and safe areas. Community leaders should also be encouraged to put injunctions on the dumping of any type of wastes (including defecating) into streams, rivers or lakes.

Many people are known to handle and use fertilizers and agro-chemicals carelessly since they do not know how dangerous some of these can be to their very health. It is obviously necessary that farmers in particular be educated on the characteristics, and trained on the proper use, of these seemingly unavoidable farm inputs.

Some water catchments are getting polluted

Water is life. Therefore, polluting water is polluting life. And since water pollution is essentially water poisoning, it also means life poisoning.

If life is valued and must be preserved, water should be treated with utmost respect and handled with the greatest care.

It is wise to avoid dumping wastes into water bodies or applying fertilisers or pesticides near the water bodies. Throwing wastes where they can be washed into water courses is also not a wise practice.

To preserve life treat water with respect and care.

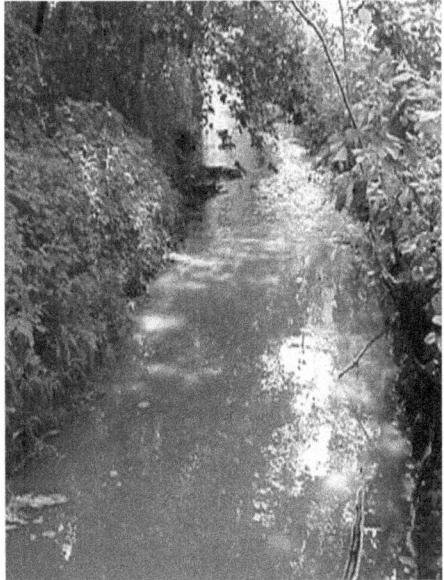
During the rainy season rivers are normally flooded and polluted from soil erosion

Source: Google

8

Water Scarcity

8.1 Definition and scale

We have learnt in Chapter 7 that about three quarters of the earth's surface is covered by water. This picture fairly describes the Bakossi Landscape. This unique landscape has a wide network of streams and rivers, as well as a collection of lakes, due to the availability of large water catchments, particularly the mountains from which the three sites derive their names. Although water is not evenly distributed, the Bakossi Landscape is so blessed that there should, naturally, be a constant supply of water for all the communities. Unfortunately, this is not the case. Why? Human-induced water scarcity!

Water scarcity describes a situation where the water supply or water availability is less than the average water requirements of a community. Although water scarcity is often defined in terms of the physical unavailability, water in available but polluted form also signifies a scarcity. This is so because it is not safe for drinking or other domestic uses. It is on record that in rural Africa about sixty five per cent of the population does not have access to adequate water supply, due to water shortages or scarcity.

It may come as a surprise that this situation seems to hold true for the Bakossi Landscape presently. But water scarcity in this landscape is increasingly becoming an important issue in some communities, with all the three sites already affected. The Bakossi site, surprisingly, is reported to be the most seriously hit, followed by the Kupe, and then by the Muanenguba site. Although the situation may not seem to be that bad in the last two sites, it is important to note that some of the communities there are already experiencing far bitter situations than the overall picture indicates. As with water pollution, this picture should serve as an early warning to the less affected sites.

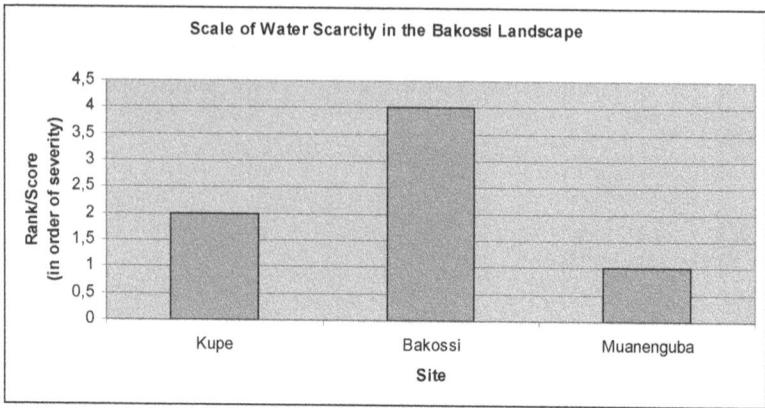

Scale of Water Scarcity in the Bakossi Landscape

8.2 Causes and underlying issues

In areas where water was once an abundant resource, scarcity can easily be ascribed to deforestation along stream or river courses and to destruction of water catchments. Besides the prevention of erosion as a factor in water scarcity, as we shall see later, it is important to note that forests also perform the important function of water conservation as the roots of some trees act as sponges, soaking, storing and releasing water gradually, which ensures a constant supply to the nearby streams, rivers, and lakes. Also, their canopies protect water catchments, streams, rivers, and lakes from direct heat of the sun and thus reduce the rate of evaporation that would have otherwise affected their water volumes.

A major issue that underlies deforestation, particularly along stream and river courses, is either outright ignorance or negligence. It is true that some people simply do not know that by clearing farms right to stream or river banks they are exposing these important resources to situations where their water volumes are considerably reduced due to silt washed into them by erosion. Consequently, these water bodies soon become shallow and may start getting completely dry during the dry season, additionally due to exposure to rapid evaporation. Ignorance may also be an underlying factor in water catchments destruction, but this situation seems to be driven more by land degradation or scarcity that eventually forces people into marginal areas, including water catchments.

8.3 Impacts and potential solutions

Water scarcity creates three main impacts in the Bakossi Landscape. First of all, the scarcity means that a lot of time that would have been

42

invested in other productive activities is spent, especially by women, searching for and fetching water.

The second impact can be seen from the low agricultural productivity gradually being observed in some areas, water being an essential requirement for agriculture.

The third impact is water-borne diseases. This situation is increasingly important, particularly in the Muanenguba site, as water scarcity correlates with water pollution in some way. In fact, water scarcity in polluted areas simply means high concentrations of pollutants, particularly the disease-causing organisms.

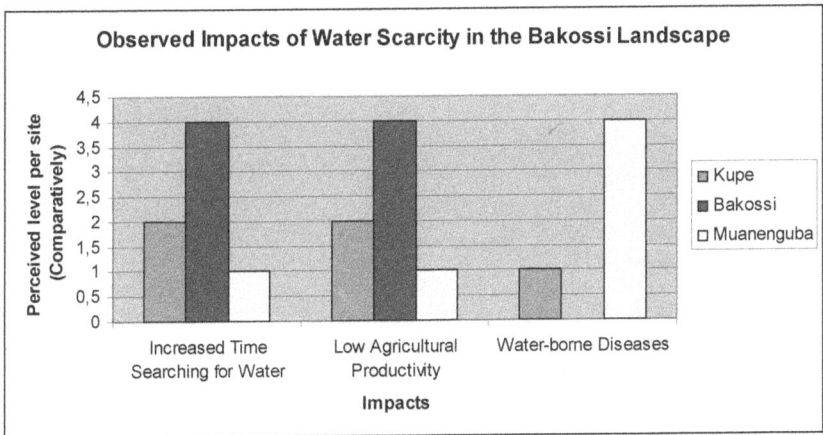

Observed Impacts of Water Scarcity in the Bakossi Landscape

Reforestation is a practical way of addressing the problem of water scarcity, given the water conservation capacity of trees as has already been discussed. Certain species of trees are important for water conservation, maintaining the water level and volume of surface water.

Identification and protection of water catchments is the surest way of preventing further water scarcity in the Bakossi Landscape. Protection efforts could start with the marking of trees bordering the water catchments with red paint as a warning against encroachment. Land use planning, culminating in the development of land use maps, is an important means of presenting a clear picture of land use patterns that would inform and support proposals for appropriate limits of land use in the area.

In all the above, community education, focusing on the causes and impacts of water scarcity, is of critical importance. This could proceed to the formation of Water Management Committees.

Natural water catchments are many in the Bakossi Landscape

Some tree roots act as a sponge

Tapping direct from natural Catchments

Some water catchments are man-made

To protect catchments, protect the hills

Queuing for water is time-consuming

We need water to live. We need water for agriculture and many domestic uses. And with water always available we have enough time to carry out our income-generating activities and fight poverty.

The Bakossi landscape is blessed with rich water catchments. But for the people to continue to live well, these areas need to be fully protected!

9

Summary and Conclusion

9.1 Key environmental problems

The Kupe and Muanenguba sites seem to carry the heaviest load of the environmental problems in the Bakossi Landscape by virtue of the fact that they experience all the seven problems, with ranks that range from low to moderately high and very high in terms of severity. The most severe environmental problems in the Kupe site are deforestation and wildlife scarcity while land degradation and species extinction are moderately severe. In the Muanenguba site, soil erosion and land degradation are the most severe while deforestation and water pollution are moderately severe. The Bakossi site is faced with four of the seven environmental problems, with deforestation and wildlife scarcity occupying very high positions and species extinction and water scarcity ranking as moderately high in terms of gravity.

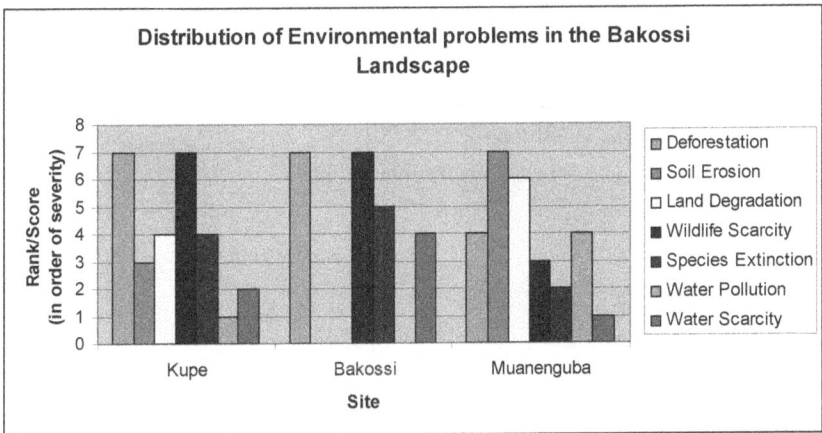

Distribution of Environmental problems in the Bakossi Landscape

After considering the total environmental load of each problem on the Bakossi Landscape (calculated by adding the total ranking by sites), we find deforestation, wildlife scarcity, species extinction, and soil erosion standing out as the key or most pressing environmental problems of the Bakossi Landscape. This classification is not only based on the total ranking or scores but also takes cognisance of the fact that by addressing the environmental problems designated as "key" we would effectively be addressing all the other problems. This explains why even land degradation which has the same score as erosion is left out.

Sites	Environmental Problems and Scores in Terms of Gravity per Site						
	Deforestation	Soil Erosion	Land Degradation	Wildlife Scarcity	Species Extinction	Water Pollution	Water Scarcity
Kupe	7	3	4	7	4	1	2
Bakossi	7	0	0	7	5	0	4
Muanenguba	4	7	6	3	2	4	1
Total Scores	18	10	10	17	11	5	7

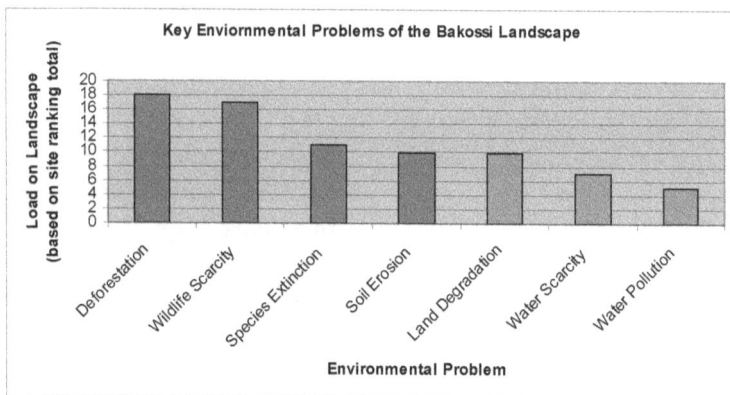

Key Enviornmental Problems of the Bakossi Landscape

9.2 Major causes and underlying issues

Environmental Problems	Majors Causes	Majors Issues
Deforestation	Shifting cultivation and logging	Poverty, ignorance, and population growth
Soil Erosion	Deforestation and over-grazing	Ignorance
Land Degradation	Deforestation, logging, and soil erosion	Poverty and ignorance
Wildlife Scarcity	Habitat destruction, over-hunting, and over-harvesting	Poverty, lack of alternative sources, commercialisation, and ignorance
Species Extinction	Habitat destruction, over-hunting, and over-harvesting	Ignorance and "I don't care attitude"
Water Pollution	Soil erosion, poor sewage and waste disposal, and inappropriate application of fertilizers	Poverty and ignorance
Water Scarcity	Water catchments destruction and deforestation	Ignorance and negligence

The Major causes and issues of the various environmental problems in the Bakossi landscape are summarised in the last table on the preceding page. Under causes, deforestation, habitat and water catchments destruction, over-exploitation, and inappropriate application of agricultural chemicals could be considered as the major direct or indirect causes of the problems. When considering the underlying issues, lack of alternative sources of protein and income and ignorance are obviously the most striking.

9.3 Main impacts and possible solutions

The table below summarises the major impacts and solutions of environmental problems in the Bakossi Landscape. In regards to the impacts already being felt, scarcity of forest products, loss of biodiversity, and scarcity of water seem to be the most important for the Bakossi landscape. The options that could be considered as the most urgent solutions are education, reforestation, water catchments protection, and promotion of alternative income and protein sources.

Environmental Problems	Majors Impacts	Majors solutions
Deforestation	Scarcity of forest products, extinction of species, scarcity of water, and increased temperatures	Environmental education and reforestation
Soil Erosion	Soil infertility, water pollution, and land degradation	Environmental education and planting of trees/grasses,
Land Degradation	Soil infertility and destruction of natural vegetation	Environmental education and planting of trees/grasses
Wildlife Scarcity	Poverty and increased time to find resources	Environmental education and promotion of alternative protein/income sources
Species Extinction	Loss of biodiversity and loss of traditional knowledge	Environmental education and law enforcement
Water Pollution	Loss of aquatic life, water-borne diseases, and sterility	Environmental education and injunctions
Water Scarcity	Increased time to get water, low agricultural productivity, and water-born diseases	Environmental education, water catchments protection and reforestation

Bibliography

Baird, Colin. 1999. *Environmental Chemistry.* New York: W.H. Freeman and Company.

Begon, Michael, Harper, John L., and Townsend, Colin R. *Ecology* (Third Edition). London: Blackwell Science Ltd.

Berküller, Klaus. 1992 (Revised Edition). *Environmental Education about the Rain Forest.* Gland: IUCN.

Brown, R. Lester, Flavin, Christopher and Postel, Sandra. 1991. *Saving the Planet: How to Shape an Environmentally Sustainable Global Economy.* New York: W.W. Norton.

Cunningham, William P., Saigo, Barbara Woodworth and Cunningham, Mary Ann. 2003. *Environmental Science: A Global Concern.* New York: McGraw-Hill.

Durning, Alan. 1992. *How Much is Enough? The Consumer Society and the Future of the Earth.* New York: W.W. Norton and Company.

Economic Commission for Africa. 2002. *Economic Impact of Environmental Degradation in Southern Africa.* Addis Ababa: Economic Commission for Africa.

Hardin, Garrett. 1998. Tragedy of the Commons. In Baden, John A. and Noonan, Douglas S. (Editors). *Managing the Commons* (Second Edition). Bloomington and Indianapolis: Indiana University Press.

Nhamo, Godwell and Inyang, Ekpe. 2011. *Framework and Tools for Environmental Management in Africa.* Dakar: CODESRIA.

Inyang, Ekpe. 2005. *Environmental Education in Theory and Practice.* Unpublished.

Middleton, Nick. 1991. *Atlas of Environmental Issues.* Oxford: Oxford University Press.

Ngome, Manasseh. 1992. *Environmental Education in Cameroon: The Problems and Prospects.* Yaounde: WWF Cameroon.

Nhamo, Godwell and Inyang, Ekpe. In Press. *Framework and Tools for Environmental Management in Africa.* Dakar: CODESRIA.

Palm, Cheryl A., Vosti, Stephen, A., Sanchez Pedro A. and Ericksen, Polly J. 2005. *Slash-and-burn Agriculture: The Search for Alternatives.* New York: Columbia University Press.

Tango Parrish Snyder (Editor-in-chief). 1985. *The Biosphere Catalogue.* Texas and London: Synergetic Press, Inc.

Acknowledgements

The writing of this handbook was inspired and largely informed by the rich outputs of the two training-of-trainers workshops organised with and for the Environmental Clubs of the Anglophone and Francophone parts of the Bakossi Landscape on March 25th and March 30th to 31st 2009 in Bangem and Loum, respectively. I would like to heartily thank the members of the Environmental Clubs and their Teacher Coordinators for their rare commitment to the improvement of the environmental situation in the Bakossi Landscape, shown through their relentless sensitisation efforts and practical measures aimed at demonstrating possible ways of addressing some of the problems.

I also extend my gratitude to the WWF Coastal Forests Programme Coordinator, Dr. Atanga Ekobo, for the support towards the realisation of this handbook, and to the Site Managers, Okon Tiku and Theophilus Ngwene, for the encouragement in the initiative of writing the book. Sincere appreciation also goes to the staff and colleagues of the WWF CFP: Yvette Mekemdem Mboneng and Jacqueline Engongwie, for the constructive comments and suggestions which helped me in no small way to improve the content of the handbook.

I would also thank Sylvie Fonkwo, a Dschang University postgraduate student on internship with the WWF Coastal Forests Programme, for her equally useful comments. Special thanks to Docaris Tene for the beautiful pictures on some aspects of the Bakossi Landscape. Thanks are also due to Dr. Christopher Agyingi of the University of Buea, for his contribution that greatly reshaped and improved the first section of the chapter on water pollution.

I will not forget to express my thankfulness to my wife, Eni, and children, Offy and Beya, for their equally critical comments and suggestions on the early draft.

Credit is also due to Google for the photos on biomass stoves on page 10 and water pollution on page 39, and to McGraw-Hill for the photo on page 20. Most of the other pictures have been obtained from the WWF Coastal Forests Programme photo library.

Appendix: Work Programme

Terms	Environmental problems	Major causes	Major Impacts	Key issues to address	Core messages	Delivery strategies	Tentative dates
First	**Chapter 2. Deforestation**	Shifting cultivation and logging	Scarcity of forest products, extinction of species, scarcity of water, and increased temperature	1. Poverty	**1.** If you cut one tree, plant two or more trees.	• Study of chapter • Campaign (Messages on placards) • Roundtable	
				2. Ignorance	**2.** Deforestation has many negative consequences that may affect us badly.	• Drama (e.g. *The Sacred Forest* by Ekpe Inyang)	
				3. Population growth	**3.** For every child you have, plant a tree each year.	• Debate • Tree nursery preparation	
	Chapter 5. Wildlife Scarcity	Habitat destruction, over-hunting, and over-harvesting	Poverty and increased time to find resources	1. Poverty	**4.** Wildlife is an important source of our income, and so we must use it sustainably.	• Study of chapter • Roundtable	
				2. Lack of alternative sources	**5.** There is no such thing as lack of alternative sources of protein and income.	• Debate	
				3. Ignorance		• Posters posted about	
				4. Commercia-lisation	**6.** Commercialisation of our wildlife destroys the very sources of our income.	• Drama (e.g. *The Game* by Ekpe Inyang)	
	Chapter 6. Species Extinction	Habitat destruction, over-hunting, and over-harvesting	Loss of biodiversity and loss of traditional knowledge	1. Ignorance	**7.** That some of our natural resources are already becoming extinct is a warning that we must be very careful the way we use the remaining ones.	• Study of chapter • Drama (e.g. *Beware* by Ekpe Inyang)	
				2. "I don't care attitude"		• Roundtable • Music presenta-tion	

51

Terms	Environmental problems	Major causes	Major impacts	Key issues to address	Core messages	Delivery strategies	Tentative dates
Second	Chapter 7. Water Pollution	Soil erosion, poor sewage and waste disposal, and inappropriate application of fertilizers	Loss of aquatic life, water-borne diseases, and sterility	1. Poverty	8. Using ago-chemicals near water courses is dangerous to our health.	• Study of chapter • Lecture	
				2. Ignorance	9. Many of the diseases in our community are due to water pollution.	• Drama (e.g. *Water na Life* by Ekpe Inyang) • Campaign (Messages on placards)	
	Chapter 8. Water Scarcity	Water catchments destruction and deforestation	Increased time to get water, low agricultural productivity, and water-born diseases	1. Ignorance	10. Our water is fast becoming scarce due to our destruction of our forest and water catchments.	• Study of chapter • Drama (e.g. *The Hill Barbers* by Ekpe Inyang)	
				2. Negligence	11. We need to do something now, or else we would face worse water crises in the future.	• Tree nursery development	
Third	Chapter 3. Soil Erosion	Deforestation and over-grazing	Soil infertility, water pollution, and land degradation	1. Ignorance	12. Our soil is our surest source of wealth and must not, therefore, be wasted.	• Study of chapter • Poster presentation • Tree or grass planting demonstration	
	Chapter 4. Land Degradation	Deforestation, logging, and soil erosion	Soil infertility and destruction of natural vegetation	1. Poverty	13. The health of our soil is our health and wealth.	• Study of chapter • Debate	
				2. Ignorance	14. Our land use pattern is fast degrading our land.	• Poster presentation • Tree planting campaign	

www.ingramcontent.com/pod-product-compliance
Lightning Source LLC
Chambersburg PA
CBHW030000290326
41935CB00008B/640